INSTRUMENT RATING KNOWLEDGE TEST GUIDE

Before procedure

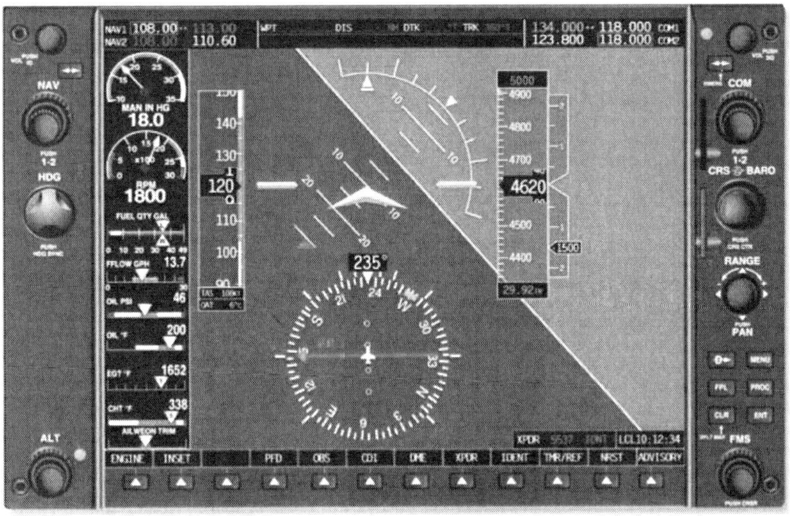

Current procedure

After procedure

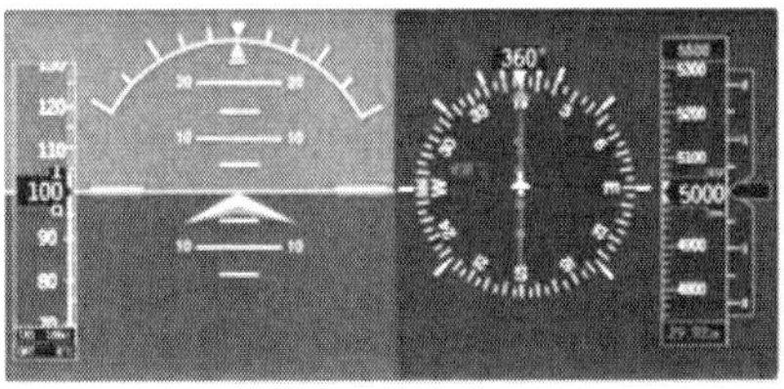

CONTENTS

INTRODUCTION

FAA-G-8082-13D, Instrument Rating Knowledge Test Guide, provides information for preparing you to take the Instrument Rating Knowledge Test. This document supersedes FAA-G-8082-13C, dated 2007.

The minimum passing score is 70 percent.

TEST NAME	TEST CODE
Instrument Rating—Airplane	IRA
Instrument Rating—Helicopter	IRH
Instrument Rating—Foreign Pilot	IFP
Flight Instructor Instrument—Airplane	FII
Flight Instructor Instrument—Helicopter	FIH
Flight Instructor Instrument—Airplane (Added Rating)	AIF
Flight Instructor Instrument—Helicopter (Added Rating)	HIF
Ground Instructor—Instrument	IGI

Federal Aviation Administration (FAA) airman knowledge tests are effective instruments for aviation safety and regulation compliance. However, these tests can only sample the vast amount of knowledge every pilot needs to operate safely in the National Airspace System (NAS).

Comments may be e-mailed to AFS630Comments@faa.gov.

KNOWLEDGE TEST ELIGIBILITY REQUIREMENTS

The general prerequisites for an instrument rating require that you have a combination of experience, knowledge, and skill. For specific information pertaining to certification, you should carefully review the appropriate sections of Title 14 of the Code of Federal Regulations (14 CFR) part 61 for instrument rating requirements.

- Additionally, to be eligible for an instrument rating, you must: Hold at least a current private pilot certificate with an aircraft rating appropriate to the instrument rating sought.

- Be able to read, speak, write, and understand the English language.

- Show satisfactory completion of ground instruction or home study course required by 14 CFR part 61 for the certificate or rating sought.

- Present as personal identification an airman certificate, driver's license, or birth certificate showing that you meet the age requirements prescribed for the certificate sought no later than 2 years from the date of application for the test.

To be eligible to take the knowledge test, you must have:

1. Received an endorsement from an authorized instructor certifying that you have accomplished a ground-training or a home-study course for the certificate or rating sought and you are prepared for the knowledge test; and
2. Proper identification at the time of application that contains your—
   a. Photograph;
   b. Signature;
   c. Date of birth, which shows you meet or will meet the age requirements of this part for the certificate sought before the expiration date of the Airman Knowledge Test Report; and
   d. Actual residential address, if different from your mailing address.

KNOWLEDGE AREAS ON THE TESTS

To be eligible to take an instrument rating knowledge test, you must have received ground instruction, or have logged home study in all the following areas:

- The 14 CFR parts that apply to flight under instrument flight rules (IFR) conditions, the Aeronautical Information Manual (AIM), and the IFR air traffic system and procedures.
- Dead reckoning appropriate to IFR navigation using various navigation systems. The use of IFR en route and approach procedure charts.
- The procurement and use of aviation weather reports and forecasts, and the elements of forecasting weather trends on the basis of that information and personal observation of weather conditions.
- The safe and efficient operation of aircraft, as appropriate, under instrument weather conditions.

The instrument flight instructor and instrument ground instructor knowledge tests cover the same subject areas listed previously. The instructor applicant is expected to have an in-depth knowledge of these areas in order to test.

DESCRIPTIONS OF THE TESTS

All test questions are the objective, multiple-choice type. Each question can be answered by the selection of a single response. Each test question is independent of other questions; therefore, a correct response to one does not depend upon, or influence, the correct response to another. **The minimum passing score is 70 percent.**

Knowledge tests for the instrument rating consist of a selection of questions in the areas that pertain to the Code of Federal Regulations (CFR) requirements, attitude instrument flying, flight planning, meteorology, the pilot's responsibility when operating under IFR; and IFR operations pertinent to preflight, departure, en route, and arrival. The Instrument Rating—Foreign pilot test includes questions that pertain to instrument flight rules and related procedures. These tests can be administered by any authorized testing center.

Ground Instructor—Instrument applicants should be prepared to answer any question that appears in the instrument question bank as they are expected to teach all instrument ratings.

The following tests each contain 60 questions, and you are allowed 2 hours and 30 minutes to complete each test.

- Instrument Rating—Airplane
- Instrument Rating—Helicopter

The following tests each contain 50 questions, and you are allowed 2 hours and 30 minutes to complete each test.

- Instrument Flight Instructor—Airplane
- Instrument Flight Instructor—Helicopter
- Ground Instructor—Instrument
- Instrument Rating—Foreign Pilot

All added instrument instruction rating tests contain 20 questions, and you are allowed 1 hour to complete each test.

TEST REGISTRATION

The FAA has hundreds of computer testing centers. These testing centers offer the full range of airman knowledge tests. List of computer testing designees (CTDs) and phone numbers are provided in the section titled Training and Testing Publications and General Information under Knowledge Test Centers.

The first step in taking a knowledge test is the registration process. You may either call the central 1-800 numbers or simply use the walk-in basis. If you choose to use the 1-800 numbers to register, you will need to select a testing center, schedule a test date, and make financial arrangements for test payment. You may register for tests several weeks in advance, and you may cancel your appointment according to the CTD's cancellation policy. If you do not follow the CTD's cancellation policies, you could be subject to a cancellation fee.

TAKING A KNOWLEDGE TEST

The next step in taking a knowledge test is providing proper identification. You should determine what knowledge test prerequisites are necessary before going to the computer testing center. Your instructor or local Flight Standards District Office (FSDO) can assist you with what documentation to take to the testing facility. Testing center personnel will not begin the test until your identification is verified.

Acceptable forms of authorization:

- A certificate of graduation or a statement of accomplishment certifying the satisfactory completion of the ground school portion of a course from an FAA-certificated pilot school.
- A certificate of graduation or a statement of accomplishment certifying the satisfactory completion of the ground school portion of a course from an agency such as a high school, college, adult education program, U.S. Armed Force, ROTC Flight Training School, or Civil Air Patrol.
- A written statement or logbook endorsement from an authorized instructor certifying that you have accomplished a ground training or home study course required for the rating sought and you are prepared for the knowledge test.
- Failed Airman Knowledge Test Report, passing Airman Knowledge Test Report, or expired Airman Knowledge Test Report (pass or fail), provided that you still have the original Airman Knowledge Test Report in your possession. Before you take the actual test, you will have the option to take a sample test. The actual test is time limited; however, you should have sufficient time to complete and review your test.

Before you take the actual test, you will have the option to take a sample test. The actual test is time limited; however, you should have sufficient time to complete and review your test.

When taking a test, keep the following points in mind:

- Carefully read the instructions given with the test.
- Answer each question in accordance with the latest regulations and guidance publications.
- Read each question carefully before looking at the answer options. You should clearly understand the problem before attempting to solve it.
- After formulating an answer, determine which answer option corresponds with your answer. The answer you choose should completely resolve the problem.

- From the answer options given, it may appear that there is more than one possible answer; however, there is only one answer that is correct and complete. The other answers are either incomplete, erroneous, or derived from popular misconceptions.

- If a certain question is difficult for you, it is best to mark it for review and proceed to the next question. After you answer the less difficult questions, return to those you marked for review and answer them. The review marking procedure will be explained to you prior to starting the test. Although the computer should alert you to unanswered questions, make sure every question has an answer recorded. This procedure will enable you to use the available time to maximum advantage.

- When solving a calculation problem, select the answer that most nearly matches your solution. The problem has been checked by various individuals and with different types of calculators; therefore, if you have solved it correctly, your answer will be closer to the correct answer than any of the other choices.

USE OF TEST AIDS AND MATERIALS

You may use aids, reference materials, and test materials within the guidelines listed below, if actual test questions or answers are not revealed. All models of aviation-oriented calculators may be used, including small electronic calculators that perform only arithmetic functions (add, subtract, multiply, and divide). Simple programmable memories, which allow addition to, subtraction from, or retrieval of one number from the memory, are permissible. Also, simple functions, such as square root and percent keys are permissible.

The following guidelines apply:

1. You may use any reference materials provided with the test. In addition, you may use scales, straightedges, protractors, plotters, navigation computers, log sheets, and electronic or mechanical calculators that are directly related to the test.
2. Manufacturer's permanently inscribed instructions on the front and back of such aids (e.g., formulas, conversions, regulations, signals, weather data, frequencies, weight-and-balance formulas) are permissible.
3. Testing centers may provide a calculator to you and/or deny use of your personal calculator based on the following limitations:
    a. Prior to, and upon completion of the test, while in the presence of the proctor, you must actuate the ON/OFF switch and perform any other function that ensures erasure of any data stored in memory circuits.
    b. The use of electronic calculators incorporating permanent or continuous type memory circuits without erasure capability is prohibited. The proctor may refuse the use of your calculator when unable to determine the calculator's erasure capability.
    c. Printouts of data must be surrendered at the completion of the test if the calculator incorporates this design feature.
    d. The use of magnetic cards, magnetic tapes, modules, computer chips, or any other device upon which pre-written programs or information related to the test can be stored and retrieved is prohibited.
    e. You are not permitted to use any booklet or manual containing instructions related to use of test aids.
4. Dictionaries are not allowed in the testing area.
5. The proctor makes the final determination relating to test materials and personal possessions you may take into the testing area.

DYSLEXIC TESTING PROCEDURES

If you are a dyslexic applicant, you may request approval from the local FSDO or International Field Office (IFO) to take airman knowledge test using one of the three options listed in preferential order:

Option 1. Use current testing facilities and procedures whenever possible.

Option 2. You may use a Franklin Speaking Wordmaster® to facilitate the testing process. The Wordmaster® is a self-contained electronic thesaurus that audibly pronounces typed in words and presents them on a display screen. It has a built-in headphone jack for private listening. The headphone feature must be used during testing to avoid disturbing others.

Option 3. If you do not choose to use the first or second option, you may request a proctor to assist in reading specific words or terms from the test questions and supplement material. In the interest of preventing compromise of the testing process, the proctor must be someone who is non-aviation oriented. The proctor must provide reading assistance only, with no explanation of words or terms. When this option is requested, the FSDO or IFO inspector must contact the Airman Testing Standards Branch (AFS-630) for assistance in selecting the test site and proctor.

Prior to approval of any option, the FSDO or IFO inspector must advise you of the regulatory certification requirement of being able to read, write, speak, and understand the English language.

CHEATING OR OTHER UNAUTHORIZED CONDUCT

Computer testing centers must follow strict security procedures to avoid test compromise. These procedures are established by the FAA and are covered in FAA Order 8080.6, (as amended), Conduct of Airman Knowledge Tests. The FAA has directed testing centers to terminate a test at any time a test proctor suspects a cheating incident has occurred. An FAA investigation will then be conducted. If the investigation determines that cheating or unauthorized conduct has occurred, then any airman certificate or rating that you hold may be revoked, and you will be prohibited for 1 year from applying for or taking any test for a certificate or rating under 14 CFR part 61.

KNOWLEDGE TEST REPORTS

Upon completion of the knowledge test, you will receive your Airman Knowledge Test Report, with the testing center's embossed seal, which reflects your score.

The Airman Knowledge Test Report lists the learning statement codes for questions answered incorrectly. The total number of learning statement codes shown on the Airman Knowledge Test Report is not necessarily an indication of the total number of questions answered incorrectly.

The Learning Statement Reference Guide for Airman Knowledge Testing, found at www.faa.gov

, contains the listings of reference materials, learning statement codes, and learning statements. The learning statement codes, as used in airman testing, refer to a measurable statement of knowledge that a student should be able to demonstrate following a defined element of training. You should match the codes on your Airman Knowledge Test Report to the codes in the Learning Statement Reference Guide to review your areas of deficiency.

Your instructor is required to provide instruction on each of the knowledge areas listed on your Airman Knowledge Test Report and to complete an endorsement of this instruction. The Airman Knowledge Test Report must be presented to the examiner prior to taking the practical test. During the oral portion of the practical test, the examiner is required to evaluate the noted areas of deficiency.

Should you require a duplicate Airman Knowledge Test Report due to loss or destruction of the original, send a signed request accompanied by a check or money order for $1.00, payable to the FAA. Send the request to:

Federal Aviation Administration

Airmen Certification Branch, AFS-760

P.O. Box 25082

Oklahoma City, OK 73125

For an Airman Knowledge Test Report to be valid, it must be dated within the 24calendar-month period preceding the month you complete the practical test. **If the Airman Knowledge Test Report expires before completion of the practical test, you must retake the knowledge test.**

RETESTING PROCEDURES

If you receive a score lower than 70 percent and wish to retest, you must present the following to testing center personnel:

- Failed Airman Knowledge Test Report
- A written endorsement from an authorized instructor certifying that additional instruction has been given, and the instructor finds you competent to pass the test

If you decide to retake the test in anticipation of a better score, you may retake the test after 30 days from the date your last test was taken. The FAA will not allow you to retake a passed test before the 30-day period has lapsed. Prior to retesting, you must give your current Airman Knowledge Test Report to the test proctor. The last test taken will reflect the official score.

TRAINING AND TESTING PUBLICATIONS AND GENERAL INFORMATION

Most of the current Flight Standards Service airman training and testing publications can be obtained in electronic format from the FAA Web site, www.faa.gov.

- The training and testing publications and general information can be found on the opening page of that Web site under the Education and Research tab. If a publication is not available in electronic format, there are instructions for obtaining paper copies. Information found on the Web site includes the following:Advisory Circulars
- Airworthiness Directives
- Code of Federal Regulations
- Computer Testing Supplements
- Knowledge Test Centers
- Knowledge Test questions
- Knowledge Test statistics
- Learning Statement Reference Guide
- Practical Test Standards
- Training handbooks
- Type Certificate Data Sheets

Advisory Circulars

Advisory circulars (ACs) provide guidance and information on various subjects related to airman certification.

Airworthiness Directives

Airworthiness Directives (ADs) are notifications to aircraft owners of a known safety deficiency with a specific model of aircraft, engine, avionics, or other system.

Code of Federal Regulations

The portion of 14 CFR containing what was formerly known as the Federal Aviation Regulations can be found on the Web site. 14 CFR contains regulations designed to promote aviation safety, and govern all aviation activities in the United States.

Computer Testing Supplements

The knowledge testing supplements contain the graphics, legends, and maps that are needed to successfully respond to certain knowledge test items. These supplements will be provided by CTD test center personnel during the airman knowledge test.

Knowledge Test Centers

The Knowledge Test Centers portion of the Web site contains current listings of CTDs and other testing centers, and the registration telephone numbers to call to register for a test.

The following is a list of the computer testing designees authorized to give FAA airman knowledge tests. This list should be helpful in case you choose to register for a test or simply want more information.

💻 Computer Assisted Testing Service (CATS)

1801 Murchison Drive, Suite 288 Burlingame, CA 94010

Applicant inquiry and test registration: 1-800-947-4228

From outside the U.S. (650) 259-8550

💻 LaserGrade Computer Testing

16821 SE McGillivray Blvd., Suite 201

Vancouver, WA 98683

Applicant inquiry and test registration: 1-800-211-2753 or 1-800-211-2754

From outside the U.S. (360) 896-9111

Knowledge Test Questions

Sample questions are located in the Airman Knowledge Test Questions section of Web site and represent the types of questions included in the actual test banks. Practicing these questions will help you become familiar with similar questions on the airman knowledge tests. The knowledge test is not designed to intimidate any prospective airman; it is designed to measure understanding of the rules and regulations required to receive an FAA certificate.

Knowledge Test Statistics

Test statistics for all airman knowledge tests are contained in a series of tables organized by year and subject area. Individual tables are provided for the following subject areas: test volume, pass rates, average test scores, countries, regions, and district offices.

Learning Statement Reference Guide

Learning statement codes replace the old subject matter codes and are noted on the test report. They refer to measurable statements of knowledge that a student should be able to demonstrate following a defined element of training. The learning statement corresponding to the learning statement code on the test report can be located in the Learning Statement Reference Guide on the Web site.

Practical Test Standards

The practical test standards outline the knowledge and skill requirements for each airman certificate and rating. The references listed in each task of the practical test standards indicate the specific publications used to develop the skill standards. The ability to issue immediate changes prior to publishing revised printed copies ensures the practical test standards are always accurate and usable.

Training Handbooks

The training handbooks are the basic information sources an airman applicant should refer to when preparing for the knowledge and practical tests for a specific certificate or rating.

Type Certificate Data Sheets (TCDS)

A Type Certificate Data Sheet contains a formal description of an aircraft, engine, or propeller, including detailed specification of the type design and the information required for type certification.

Notes:

Notes:

Notes:

Notes:

Notes:

Notes:

Notes:

Made in United States
Cleveland, OH
06 November 2026

47782094R00015